© Aladdin Books Ltd 2001

Designed and produced by
Aladdin Books Ltd
28 Percy Street
London W1P 0LD

*First published in
the United States in 2001 by*
Copper Beech Books,
an imprint of
The Millbrook Press
2 Old New Milford Road
Brookfield, Connecticut 06804

ISBN 0-7613-2459-3 (lib. bdg.)
ISBN 0-7613-2335-X (paper ed.)

*Cataloging-in-Publication data is on
file at the Library of Congress*

*Printed in Belgium
All rights reserved*

Coordinator
Jim Pipe

Design
Flick, Book Design and Graphics

Picture Research
Brian Hunter Smart

Illustration
Mary Lonsdale for SGA

Picture Credits
Abbreviations: t – top, m – middle,
b – bottom, r – right, l – left, c – center.
All photographs supplied by
Select Pictures except for:
Cover, 24bl – Corbis. 2tl, 23br – Stockbyte.
3, 22tl – Gerard Lacz/FLPA-Images of
Nature. 6tl, 17, 23tl – Digital Stock. 7, 15,
20-21, 22tr – Corbis/Royalty Free. 9, 10tl,
24br – Scania. 12tl – Jacques M. Chenet
/CORBIS. 18b, 23mr – Neil Rabinowitz
/CORBIS. 24tr – John Foxx Images.

Sounds

By Dr. Alvin Granowsky

Copper Beech Books
Brookfield, Connecticut

 # Quiet

Kevin and Jill are waiting for the parade. What sounds can they hear?

It is quiet. They hear the wind in the trees and a bird singing.

2

A mouse is very quiet.

Can you be as quiet as a mouse?

Loud

Bang, bang, bang!

What is that sound?

It is a hammer. It is very loud.

It makes the bird fly away.

Brrm, brrm! What is that sound?

It is a motorcycle.

Far away, it sounds quiet.

Close up, it sounds very loud.

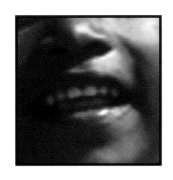

Talking

What is that sound?

It is Kevin talking. But Jill wants to listen for the parade. She puts a finger to her lips. Shh!

When you talk, you make sounds into words. In writing, letters stand for sounds.

Signals

Ding-dong!
What is that sound?

It is the bell ringing.
It is 11 o'clock.

Time for
the parade!

Some sounds make you jump.

An alarm clock says, "Wake up."

A siren says, "Watch out!"

Machines

Brrm! Brrm!

What is that sound?

Here are the trucks in the parade.

Their engines rumble.

Some machines are noisy.

Others hum quietly.

This machine is very noisy.

Cover your ears!

Voices

What is that sound?

Someone is singing!
The crowd shouts
and cheers.

Your voice can make many sounds,
high or low, loud or soft.

You can shout or scream,
giggle or hum!

One, two, three

What is that sound?

The parade marches by.

The feet go tramp, tramp, tramp.

Everyone claps.

When you clap, your hands make a sound. Clap, clap, clap!

This bird makes a sound with its beak. Tap, tap, tap!

High and low

Boom! Boom! What is that sound?

It is a big drum.
It makes a
low sound.

A triangle goes
ding, ding!
It makes a
high sound.

ROAR!

A bear's roar is a low sound.

You can hear it far away.

Music

Da, da, da-da!

What is that sound?

The trumpets are playing music.
The band is very loud!

Music can make you feel happy.

It can make you want to march
or dance!

Bang!

What is that sound?

Fireworks.

Fireworks make a lot of noise.

Some fireworks whistle.

Some go whoosh!

And some just go BANG!

Here are some words about sound.

Quiet

Loud

High

Noisy

Low

These things make a sound.

Voice

Hammer

Hands

Band

Can you write
a story with
these words?

Bird

Do you know?

You can write down many sounds.

A clock goes
tick tock.

A rooster goes
cock-a-doodle-doo!

Water goes
splash!

What sound does
a fire engine make?

24